# Class 442s
## The Wessex Electrics

MARK V. PIKE

BRITAIN'S RAILWAYS SERIES, VOLUME 27

**Front cover image**: 2409 is seen on the approach to Weymouth, just about to pass under the Alexander Footbridge. This was the location of countless images taken during steam days when Weymouth Motive Power Depot used to be located where the housing estate is, just to the right above the train. The observer back in the day would be treated to many engine movements from the depot to and from the station and yards. 15 October 1994.

**Back cover image**: Two unidentified 442s are seen at Eastleigh station on a fine early winter morning. 9 December 2003.

**Title page image**: First and last at Poole! 2401 and 2424 are seen waiting to depart Poole with a London-bound service. Notice the differing liveries, with the revised version on 2401 and the original on 2424. 4 December 2006.

Published by Key Books
An imprint of Key Publishing Ltd
PO Box 100
Stamford
Lincs PE19 1XQ

www.keypublishing.com

The right of Mark V. Pike to be identified as the author of this book has been asserted in accordance with the Copyright, Designs and Patents Act 1988 Sections 77 and 78.

Copyright © Mark V. Pike, 2021

ISBN 978 1 80282 184 0

All rights reserved. Reproduction in whole or in part in any form whatsoever or by any means is strictly prohibited without the prior permission of the Publisher.

Typeset by SJmagic DESIGN SERVICES, India.

# Introduction

Based on the very successful Mk3 coach design, similar to the HST, the first of the new futuristic looking Class 442s was officially handed over to Network SouthEast on 18 December 1987 at Derby Litchurch Lane, where the units were constructed. The first electric multiple unit capable of 100mph, it made its way behind a diesel loco to Bournemouth Traction and Rolling Stock Maintenance Depot (TRSMD), where they were all to be based at the end of January 1988. At first, they were solely intended for use on the Waterloo to Weymouth line, although they later spread their wings to operate on the London Waterloo to Portsmouth Direct line in the mid-1990s. After the privatisation of British Rail in the late 1990s, all the units were transferred to Angel Trains and were then leased by South West Trains. They continued work on fast and semi-fast services to Portsmouth and Weymouth from Waterloo and all eventually received South West Trains' stylish new livery during the 2000s. All bar three of the class received names, mostly early on in their career, and most in keeping with the area they served. All these names were removed, however, after withdrawal by South West Trains and were auctioned off for charity some years later. Neither Southern nor South Western Railway ever bestowed any of the units with names. Units 2401+2403 still hold the world speed record for a third rail multiple unit of 109mph, which was achieved during a special non-stop run from Waterloo to Weymouth in 1988, which they completed in just under two hours.

After the class suffered a few failures and were becoming a bit more unreliable, South West Trains decided to go for a whole new fleet of Class 444/450 'Desiro' units. The Class 444s were a direct replacement for the 442s, which again found themselves running on the Weymouth line only for a

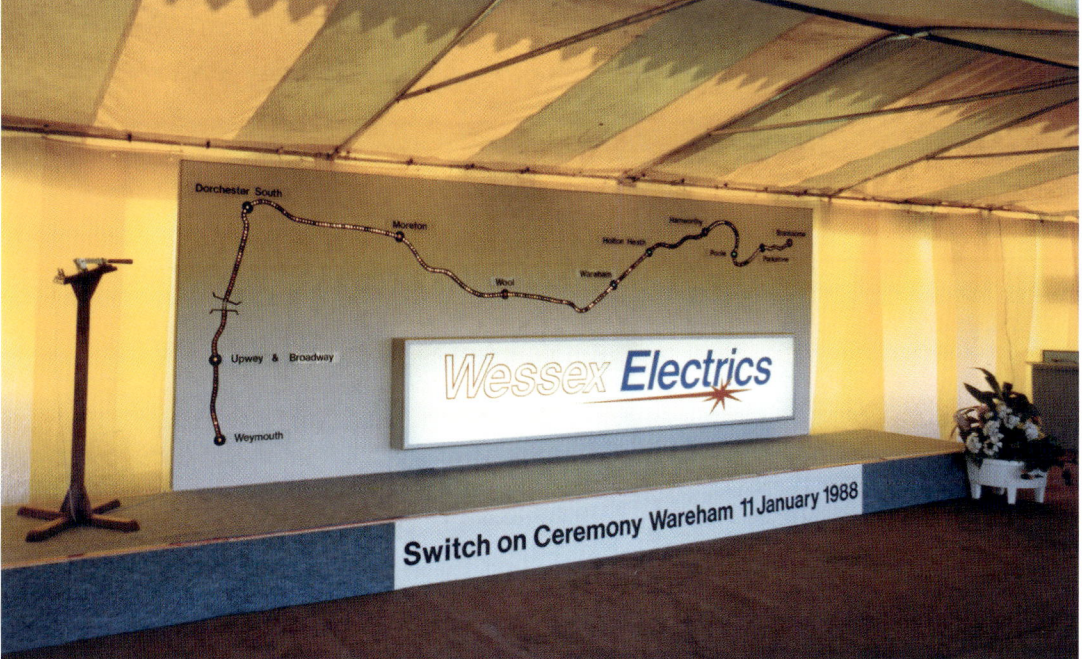

short period. However, although the lease had ended, it was still something of a surprise when South West Trains withdrew them from service in early February 2007, even more so as they were halfway through a refurbishment programme.

Seemingly too good for scrapping, 17 Class 442 units later found employment on Southern's Gatwick Express, which commenced in December 2008, initially working alongside their Class 460 units on non-stop services from London Victoria to Gatwick Airport, with some services later being extended to Brighton. Eventually, more 442s were taken on by Southern and the Class 460s were removed from service to be modified within the Class 458 units working for South West Trains. However, things never remain the same for long, and on today's railway and by mid-2016, Gatwick Express had a brand-new fleet of Class 387/2 units to replace the ageing Class 442s, which were then withdrawn again, this time ending up in storage at Ely Papworth Sidings, of all places.

However, this was not the end of the saga as one might expect! Not long after being withdrawn from Gatwick Express/Brighton services in March 2017, South Western Railway, which by then had taken over the former South West Trains franchise, had announced it was to take on 18 units, along with refurbishment and repainting, to work on London to Portsmouth direct services. This re-entry, however, would not commence until May 2019. Continuous issues dogged the units, such as door lock problems and supposedly turning signals back to red, which had resulted in the temporary removal of the units for rectification.

All faults now seemingly ironed out, early January 2020 saw another entry into service for South Western Railway, but this was only to last a couple of months until the COVID-19 pandemic took hold. Subsequent travel reduction and restrictions once again caused all sets to be withdrawn from service in March 2020. The units were then stored once again.

Then, in March 2021, South Western Railway announced that the Class 442s will no longer undergo any further refurbishment and would be returned to Angel Trains. In light of the huge amount of money that had been spent on the fleet (even including re-tractioning a few), this was a big surprise and what seemed like an awful waste of money. The only thing left for them now, it seems, is the scrapyard, and indeed by October 2021, a few units have already succumbed.

**2413 is seen crossing part of Poole Harbour as it approaches Poole with a Weymouth to London Waterloo train. 3 December 2004.**

# Class 442s

*4422401 Beaulieu*

Seen inside the new purpose-built shed at Bournemouth, this is brand-new unit 2401 receiving the attention of various fitters. The unit was delivered by diesel loco on 29 January 1988, yet to be fully fitted out internally whilst it was used for various tests. It made its first run under its own power on 11 February of the same year. 17 February 1988.

Seen here at the very end of the line that it was designed for, 2401 has just arrived at Weymouth with a service from London Waterloo. 25 March 2003.

2401 is captured here about a mile east of Dorchester South at Syward, with a Weymouth to London Waterloo service. The train is approaching a footpath over the line, which was originally a double-track road crossing. The relatively new bridge, prominent above the train, is the Dorchester bypass, constructed in the 1990s. The brick built one in the distance has the wonderful name of Smokey Hole Bridge, which, back in the days of steam, was probably very accurate! 2 December 2006.

This is Wareham station on a damp early spring morning; formerly the changing point for the branch line to Swanage, it indeed should be once again in the not-too-distant future. The unit is waiting to depart with a Weymouth to London Waterloo service. 9 March 2005.

The same train as the previous image, seen passing the now redundant (since the mid-2010s) signalbox. The crossing in the foreground has been very controversial in recent years and even featured in the national news. It has the unenviable reputation of being cited as the 'most dangerous' on the rail network. It is a little ironic that it was originally a fully gated road crossing before the construction of the bridge seen in the background. Network Rail has come up with many different plans to replace it, but all have been rejected to date. One they have never come up with is an underpass, which to my mind would solve all their problems! 9 March 2005.

Here we see the first of a few shots taken on the same day in March 2005. 2401 is pausing beneath the massive overall roof of Bournemouth station with a Wareham-bound train. Bournemouth station was opened in 1885, first as Bournemouth East, then Bournemouth Central and eventually just Bournemouth from 1967 when the other station in the area, Bournemouth West, was closed. It originally had two through lines, hence the wide spacing between the tracks. The magnificent roof fell into neglect for many years and was further affected by the storm of 1987, when many glass panels were blown out, presenting a real danger to the travelling public. In 2000, it was finally restored to its former glory and still looks splendid to this day. The only part that spoils the whole thing is the awful modern-looking footbridge (seen in this image) that was installed in the 1970s and looks totally out of place. It originally replaced a much more sympathetic ramped subway (which has since returned to use, ironically) that passes underneath the platforms, around where the fourth coach of the train is standing. 11 March 2005.

2401 spent most of that same day working on various Brockenhurst to Wareham/Weymouth stopping services that operated at the time and is seen arriving at Parkstone station heading for Wareham. The steep rise at 1–50 commencing the other side of the ornate bridge is very apparent, and even for modern electric units, pulling away from a station stop here can be very awkward, especially in wet conditions. 11 March 2005.

A little later in the day, the unit is seen here approaching Parkstone again, this time from the Poole direction and coming up the two miles or so of a rising 1–60 gradient. Parkstone station itself is actually situated on a much flatter gradient, but the line rises at 1–60 from the south and changes again to 1–50 just past the bridge mentioned in the previous image. 11 March 2005.

This 2401 is pausing for custom at Parkstone station before heading north, towards Branksome and Bournemouth. The South West Trains stylish 'swoosh' livery can be seen to good effect. 11 March 2005.

In the mid-2000s, Brockenhurst to Wareham stopping services invariably used the up loop platform at Brockenhurst station, and this is where 2401 is seen soon after arriving from Wareham. 11 March 2005.

Later, 2401 is seen at the end of the line at Weymouth, having just arrived from Brockenhurst. 11 March 2005.

The last shot taken on this day sees the unit once again under the overall roof at Bournemouth, but this time in the up platform, waiting to depart for Brockenhurst. The aforementioned subway goes under the line behind the low wall with the poster on, beside the cab of 2401. 11 March 2005.

Approaching Wareham with a Waterloo to Weymouth service, 2401 passes one of its successors in the shape of 444023, which is lurking in the sidings. 2 January 2007.

## Class 442s

Gliding to a stop on the approach to Moreton station with a Weymouth to Waterloo service. The unit still looks very smart, despite it being just a few weeks until withdrawal by South West Trains. 2 January 2007.

The same service as seen in the previous picture, with the unit departing Moreton and passing over the level crossing. This has been another quite dangerous crossing over the years. It is actually situated in a hidden dip in a straight, fast road, where many an idiot driver has driven around the half barriers. 2 January 2007.

After ceasing work with South West Trains, 2401 is seen here at Lewes with a crew/route learning run to Eastbourne. Although mainly used on Gatwick Express services, there were a couple of diagrams to Eastbourne. 18 February 2010.

*Above*: On the same very wet day as the previous image, 2401 is seen approaching Eastbourne station. 18 February 2010.

*Left*: When these units were first put into service, I do not think anyone would have expected to see them on the stops at Eastbourne, of all places! 2401 rests soon after arrival. Regrettably, this unit was scrapped during 2020, but at least one of the driving coaches was saved and now, rather bizarrely, at the time of writing, resides in Northumbria! 18 February 2010.

## 4422402 *County of Hampshire*

## Class 442s

The first of two images showing 2402 virtually brand new, stabled at Bournemouth West TRSMD. This unit, as 2401, was used for a few months for driver training/testing and was delivered without many of its internal fittings. 26 March 1988.

This image clearly shows the small portion of open second-class accommodation, situated right at the front of the first-class corridor section. This was quickly altered to be open first seating, in line with the first-class accommodation of the rest of the coach. As far as I am aware, none of the units ever ran in service in the configuration pictured here. 26 March 1988.

This is 2402 approaching Eastleigh non-stop with a London Waterloo to Weymouth service. The units suffered from a well-known fault, in that the front jumper cable covers would never seem to stay in place. They looked fine when new (see previous image), but after a while in use they often did not go back properly or fell off altogether! There was a long period of use where they were actually removed, mainly through frustration, I think! 25 March 2003.

## Class 442s

A chance passing shot of 2402 on a Waterloo-bound service and 2418 on a Bournemouth service, captured at Mount Pleasant Crossing, just south of St Denys. 11 October 2004.

During its time working for Southern and Gatwick Express services, 2402 is seen crossing over the line to Kensington and Willesden at Culvert Road, near Clapham Junction, with a Victoria to Gatwick Airport service. 13 December 2014.

This is 2402 leading another unit on the approach to Clapham Junction with another Victoria to Gatwick and Brighton service. Most of these trains were originally non-stop between the two points, but eventually stops were made at Clapham Junction and East Croydon, with some services extending to Brighton. At the same time those stops were added, in 2012, the branding on the trains was altered to just 'Express' rather than 'Gatwick Express', as not all of these stopped at Gatwick, and so it could be a little misleading. 9 July 2016.

## Class 442s

### 4422403 *The New Forest*

# THE NEW FOREST

2403 is seen here passing through Farnborough (Main) with a Weymouth to London Waterloo service. This unit was readily identifiable at the time for the somewhat deformed corridor connection on one end! I am not sure what happened – perhaps it had a bit of an over enthusiastic shunt at some point? 12 December 2006.

This is the approach to Dorchester South station with 2403 on a service for Weymouth. This was originally a double tracked mainline, but the section between Dorchester South and Moreton (around seven miles or so) was singled back in the mid-1980s, supposedly as an economy measure, as much of the track needed replacing at the time. However, rather than renew it, they just took it away and never replaced it. The long siding to the right was actually part of the former down line. 8 March 2005.

This time we see the unit arriving at Poole with a Weymouth to London Waterloo service. Although a couple of sidings do remain, the area to the right was once a substantial goods yard that was busy both day and night in its heyday. 21 October 2006.

Seen again at Poole, this time during its last two weeks of service for South West Trains, 2403 is waiting to leave with a Poole to London Waterloo semi-fast service. It can be appreciated here just how tight a curve the station is situated on. 2 January 2007.

After some time in storage at Eastleigh, whilst the powers that be decided on their future, 2403 worked various 'warming' runs from Eastleigh before entering service with Southern at the end of 2008. It is seen here passing through a rather damp South Croydon with a Brighton to London Victoria service. 14 October 2014.

After Southern dispensed with them in the mid-2010s, things once again looked bleak for the fleet, and much toing and froing took place whilst their future was mulled over once again. 2403 was one of the class that finally came full circle back to where they started life in 1987 and was taken on by South Western Railway (the successor to South West Trains) and repainted in its latest livery. This is the unit passing Southampton Central with a Fratton Depot to Bournemouth TRSMD crew familiarisation run. Personally, I think this livery quite suited the class, almost looking like a brand-new train. 24 April 2019.

At the start of 2020, the future for the class looked very positive, with some units entering service with South Western Railway on the Portsmouth Direct line via Guildford. This is 2403 on the rear (with 2408 leading) on a midday service to London Waterloo departing the Woking stop. This situation all changed, of course, with the unprecedented arrival of the COVID-19 pandemic in March of that year. 7 January 2020.

## Class 442s

### 4422404 *Borough of Woking*

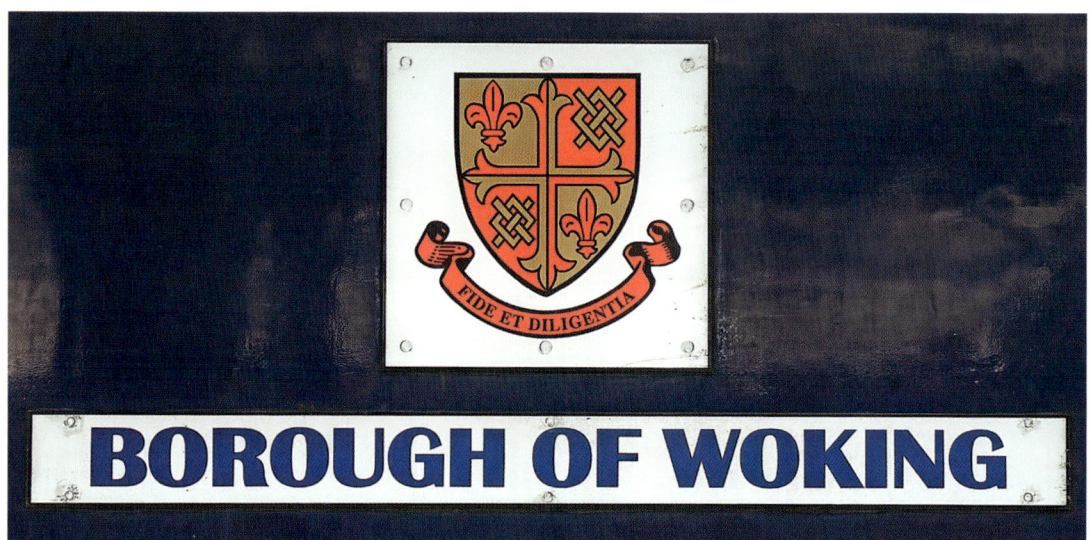

This is 2404 departing Southampton Central with a semi-fast London Waterloo to Poole service. 4 September 2003.

With the brakes applied for the stop at Winchester, 2404 is working a London Waterloo to Weymouth service. The bridge, seen here above the train, has since been replaced with a new concrete structure to ensure clearance for larger intermodal container boxes. 6 November 2006.

*Above*: Passing a field of winter fodder at West Stafford on the outskirts of Dorchester, 2404 is heading for London Waterloo on the single-line section to Moreton with a service from Weymouth. The grand looking house on the top right of the image is the agricultural college at Kingston Maurward. 2 December 2006.

*Right*: Here we see 2404 picking up passengers at Basingstoke whilst working a Poole to London Waterloo semi-fast service. 5 December 2006.

Here is 2404 departing Southampton Central, about to enter Southampton Tunnel with a Weymouth to London Waterloo train. 6 December 2006.

South West Trains had already withdrawn many units from service before this image was taken of 2404 at Branksome. On the very day it was withdrawn, the unit is working from Bournemouth TRSMD to Eastleigh for further storage, along with 2406, which is out of sight past the bridge. 15 January 2007.

Eventually, 2404 was another unit taken on by Southern and is seen here approaching East Croydon with a Brighton and Gatwick Airport to London Victoria service. 27 January 2012.

The unit's former stamping ground of the South Western Main Line is to the right of the picture here as 2404 comes off the line from Croydon on the approach to Clapham Junction with a Brighton to London Victoria train. 22 October 2014.

There was a short period of time when a couple of units were painted in a basic South Western Railway livery, but without any decals etc. This is the ghostly looking 2404, bringing up the rear of a Fratton Depot to Bournemouth TRSMD crew-training run. This view is impossible now that the bridge on which I was standing has been removed. 18 October 2018.

## 4422405 *City of Portsmouth*

2405 is seen making its scheduled call at Winchester with a Weymouth to London Waterloo train. 27 March 2003.

In some fine early spring sunshine, 2405 is seen pulling away from Parkstone (Dorset) with a Waterloo-bound semi-fast service from Poole. Like most stations, this one had a small goods yard, which was situated to the right of the picture behind the hedge, and there was even a short goods line from here that lead to a large pottery complex, but this was closed in 1967. 11 March 2005.

A very clean 2405 is seen passing the Freightliner terminal at Millbrook (Hants) with a London Waterloo-bound train. 7 June 2006.

Taken on the same day as the previous image, this time 2405 is approaching Wareham station with a Weymouth-bound train. The train is passing over what was originally a road crossing, replaced in the 1980s by the road bridge seen in the background. At this time, the signalbox on the left was still in operation but has since been closed, although it does still stand. The old goods shed in the background looks to be on a strange angle to the railway, but it was originally part of the first station at Wareham, which was situated in this vicinity from 1847 to 1887. It is a listed building and has since been restored, now housing some unusual office accommodation. 7 June 2006.

Here we see 2405 heading west and approaching the site of the former Lymington Junction, about a mile west of Brockenhurst in the New Forest National Park. The line to the right is the single track to Lymington Pier, which did not exist prior to the mid-1970s when the actual junction (just behind the bridge I was standing on) was dispensed with. 2 August 2006.

Further east along the South West Main Line, the unit is approaching Basingstoke with a London Waterloo-bound train. 5 December 2006.

Further west once again, 2405 is seen departing Wool with a Weymouth to London Waterloo service. In the background, Freightliner's 66622 can be seen as it waits to depart as soon as 2405 leaves the section with a sand train for Neasden. These sand trains have since finished, but there have been various rumours over the years that they will start up again; nothing has come of it to date, however. 6 December 2006.

Just 12 days before the final day the 442s worked for South West Trains, 2405 is waiting to leave Poole with a train for Weymouth. 22 January 2007.

2405 was one of just three units in service on the 442s' final day with South West Trains (3 February 2007) and was another unit to find further work for Southern. The unit is seen here approaching South Croydon with a London Victoria to Brighton service. 24 July 2012.

The unit is again seen passing South Croydon, on the same day as the last image, with a return Brighton to London Victoria train. 24 July 2012.

Another London Victoria to Brighton service is seen passing East Croydon. The units rarely deviated from this route whilst working for Gatwick Express. 27 August 2014.

## Class 442s

### 4422406 *Victory*

**VICTORY**

*Above*: Passing through St Denys station, 2406 is working a Poole to London Waterloo service. The footbridge in the background had long been out of use by this date and was removed very soon after this picture was taken. 4 December 2003.

*Right*: Soon after getting away from Southampton Central, the unit is passing what would eventually become South West Trains', and more recently South Western Railway's, Northam Depot (out of shot to the left) with a Weymouth to London Waterloo train. 11 October 2004.

Further to the mention of Lymington Junction, this is 2406 passing the actual site of what was once the three-way junction heading for London Waterloo. Originally, there was a signalbox roughly to the left of the rear coach, with the single-line Lymington branch coming off of the down main line and away to the left, and the original double track main line to Ringwood, Wimborne, Broadstone and Poole heading off to the right above the last coach. 2 August 2006.

2406 (with 2404) is seen approaching the main line at Branksome, coming off the short line from Bournemouth TRSMD and heading for storage at Eastleigh Works on the very day it was taken out of service by South West Trains (see also p 20). Many years ago, the area behind the train was the location of the small Branksome loco depot in steam days, which was a signing on and off point for steam crew working the long-closed Somerset & Dorset Railway, and further on past what is now Bournemouth TRSMD was the terminus station used by these Somerset & Dorset services at Bournemouth West, which closed in 1965. 15 January 2007.

Here we see the unit on a test run from Bournemouth TRSMD to Southampton Up Goods Loop, passing through Millbrook (Hants). This was another 'warming' run, performed just prior to it being taken on by South Western Railway. 2 August 2018.

During the time the units were initially in service with South Western Railway in 2019, they were used on a few services on the Bournemouth line to/from Poole but were twice withdrawn, once due to door defects and again because they were thought to be turning certain signals to red on the South Western Main Line. These problems were finally rectified towards the end of 2019, and it was then the intention for them to be used mainly on the Portsmouth direct line to/from Waterloo via Guildford from early 2020. As they were based at Bournemouth TRSMD, there were, however, a couple of early morning and late evening diagrams to get them back for maintenance. This is 2406 approaching Southampton Central with a morning Poole to London Waterloo train. 24 July 2019.

Class 442s

**4422407** *Thomas Hardy*

# THOMAS HARDY

*Above*: With Southampton Central station just glimpsed below the signal gantry, 2407 heads south past Freemantle footbridge with a London Waterloo to Weymouth service. 11 July 2006.

*Left*: A blustery day at Poole sees 2407 arriving with a Weymouth to London Waterloo train. The floodlights in the background are located in Poole Stadium. 29 November 2006.

The large space between the platforms at Pokesdown is testament that four tracks were once in place here, but these were removed at the end of steam operations in 1967. In fact, many years ago, Pokesdown was actually built as an island station with tracks running either side of it. 2407 is on the rear of a London Waterloo to Poole service pulling away from the station stop, just a month before the end of its use with South West Trains. 3 January 2007.

More infrastructure that has since been removed, as the unit arrives at Hamworthy on a damp winter day with a Weymouth-bound service. The signalbox here lasted in use until demolition came in spring 2014 upon re-signalling of the line. However, it was downgraded in importance many years before this. Hamworthy was also the junction for the two-and-a-half-mile line to the first Poole station; the track is still extant to this day, being used for freight only (though effectively now mothballed in 2021) for an amazing 125 years, with the old station closing to passengers in 1896! It was also the junction for the original line via Wimborne and Ringwood to Brockenhurst, often known as 'The Old Road', which closed in 1964. 13 January 2007.

2407 is now waiting to depart Hamworthy with the service seen in the previous picture. Until fairly recently, this was the only remaining semaphore signal on the whole South West Main Line, although it had actually long been disconnected. It was finally removed in May 2014 as part of the aforementioned signalling upgrade scheme. 13 January 2007.

2407 was withdrawn by South West Trains at the end of January 2007, and after a time in storage, the unit was transferred to Southern during 2008. It is seen here on a driver training run at Three Bridges. 13 November 2008.

When first working for Southern, most of the Gatwick Express services were formed of two Class 442s, especially at peak times. However, various factors, including failures and maintenance issues, lead to just solo units operating. This is 2407 approaching Clapham Junction with a London Victoria to Gatwick Airport and Brighton service. 13 December 2014.

**4422408** *County of Dorset*

This is 2408 drifting down the incline at Baiter, between Parkstone and Poole, with a Weymouth-bound train. I took this unusual view from a boat in Poole Harbour! The area of land in the foreground was reclaimed during the 1970s – the sea originally went right up to the base of the embankment which, of course, was then free of any undergrowth. 12 November 2003.

Racing south through Micheldever this time, the unit is heading for Weymouth with a service from London Waterloo. Regrettably, the bridge from which this image was taken was reconstructed not long after the date of this image, with much higher parapets, so unless one is about 7ft tall, some sort of item to stand on would be required here these days. 15 April 2003.

A very smart looking 2408 is seen here at the end of the line at Weymouth, waiting to form another service for London Waterloo. 29 September 2006.

When in service with South West Trains, the units were sometimes diverted from their usual routes due to engineering work, mostly at weekends. This is 2408+2414 on the approach to Buriton Tunnel, near Petersfield, with a London Waterloo to Weymouth train. 9 December 2006.

Taken out of service with South West Trains in mid-January 2007, this was another example to see further service with Southern and is seen here approaching East Croydon with a Brighton and Gatwick Airport to London Victoria train. 27 August 2014.

The unit is seen back on its old stamping ground, approaching Southampton Central in pouring rain with a Fratton Depot to Bournemouth TRSMD crew training run. 24 April 2019.

The return to old haunts is now complete as 2408 calls at Woking with a Portsmouth to London Waterloo service. Little did anyone know at the time, but the devastating COVID-19 pandemic that took hold soon after the date of this image would change just about everything, and the 442s were withdrawn for the very last time. 7 January 2020.

## 4422409 *Bournemouth Orchestras*

The original Network Southeast livery looked very stylish when the units were new and suited them very well. 2409 is captured drifting down the grade from Bincombe Tunnel (between Dorchester and Weymouth) with a Weymouth to London Waterloo service. 22 June 1993.

Seen again in original livery, 2409 is on the approach to Weymouth and just about to pass under the Alexander Road footbridge. This was the location of countless images taken during steam days, when Weymouth Motive Power Depot used to be located where the housing estate is, just to the right above the train. The observer back in the day would be treated to many engine movements to and from the station and yards. 15 October 1994.

**The South West Trains 'swoosh' livery is seen to good effect here as 2409 pauses for custom at Pokesdown with a Poole to London Waterloo semi-fast service. 15 September 2004.**

**Since 2014, Dorchester South signalbox has been the only one still in operation between Bournemouth and Weymouth, as it controls the line from Dorchester to Weymouth and up to Maiden Newton on the Bristol line. Eventually withdrawn by South West Trains in early January 2007, this is 2409 passing by the box with a Weymouth to London Waterloo train. Note that at least one unit's nameplate was missing by this time. 11 March 2005.**

## 4422410 *Meridian Tonight*

This unit was one of the very last to work for South West Trains, seeing quite a lot of use in the last few months or so. Here we see it arriving at Poole with a London Waterloo to Weymouth service. 4 December 2006.

Catching a glint from the low winter sunshine, this is 2410 working with 2403 at Finchdean, near Rowlands Castle, with a diverted Weymouth to London Waterloo service. This was one of the last times the class were seen on diverted runs on this route before withdrawal by South West Trains. 9 December 2006.

Occasionally, the units saw use on the Brockenhurst to Wareham stopping services, especially during their last few months of service with South West Trains. This is 2410 in the sidings at Wareham during its short layover before heading back to the station to work another stopper to Brockenhurst. 13 January 2007.

With just a couple of weeks to go before withdrawal, 2410 is seen in the bright winter sunlight as it calls at Winchester with a London Waterloo to Weymouth train. 17 January 2007.

After taking the previous image, I boarded the train at Winchester for the trip to the end of the line. It is seen soon after arrival at Weymouth, where I secured a further shot of it. 17 January 2007.

The next two images were taken under the superbly restored overall roof at Bournemouth station, another location that once had two centre roads. There was also a large motive power depot here back in the days of steam, the whole place being a hive of activity. 2410 waits to depart with a London Waterloo to Weymouth train. 24 January 2007.

*Left*: When the station at Bournemouth was refurbished back in 2000, there was talk of the platforms being widened and tracks slewed to close up the wide space in between. However, this came to nothing, and it remains the same to this day. 24 January 2007.

*Below*: The next three images are taken on the penultimate day of Class 442 operation with South West Trains. On the embankment just west of the former Monkton and Came Halt, 2410 is on the final few miles to its destination with a London Waterloo to Weymouth service. 2 February 2007.

After taking the last image, it was down to Weymouth for another shot. The unit is seen at rest under the remaining original part of the station on Platform 3. 2 February 2007.

This historic image shows what is thought to be the very last daylight departure of a service train using a Class 442 from Weymouth, the 16.00 to London Waterloo. I believe there was one later departure to Bournemouth, but this was after dark. On the final day (Saturday 3 February), none of the remaining three units, 2405, 2410 and 2412, actually worked to Weymouth. 2 February 2007.

After some time in storage at Eastleigh Works, 2410+2406 are seen passing the Baltic Siding on the approach to Winchester, returning from Basingstoke after an out-and-back 'warming' run from the works. This was operated by Freightliner Heavy Haul drivers to make sure all was working properly, prior to a move to Selhurst Depot during 2008. 17 May 2007.

2410+2401 are seen at Eastbourne after arrival on a test run from Selhurst TRSMD to check passenger elements of operation prior to the introduction of the units on Eastbourne to London Bridge peak workings. 18 February 2010.

*Above left*: Here is a look at the first-class seating on 2410 when it worked for South West Trains.

*Above right*: The second-class seating arrangements.

*Below left and below right*: An example of the 'snug' area near to the buffet in the second-class section. This whole area was redesigned after South West Trains finished with them.

Fast forward another eight years, and 2410 is seen back on the South Western Main Line, this time with a Bournemouth TRSMD to Southampton Up Goods Loop test run. 2410 was retained for a while by Southern and used on Eastbourne services, after its use on Gatwick/Brighton services were taken over by Class 387s. It was finally withdrawn by Southern in March 2017 and refurbished for South Western Railway during 2019. 2 August 2018.

Now looking very smart in the latest South Western Railway livery, 2410 is seen approaching Eastleigh with a transfer move from Bournemouth TRSMD to Eastleigh Works, unusually under its own power. At this time, many of these sorts of moves were hauled by Rail Operations Group (ROG) diesel locos. 4 March 2019.

During the period in 2019, when some units were being used on an early morning service from Poole to London Waterloo, the units were then stabled in Clapham Yard all day until the evening peak, when they worked back to Bournemouth and Poole. 2410 is seen in the carriage shed at Clapham Junction. Unfortunately, since the en masse withdrawal of these units by South Western Railway in mid-2020, the unit has since been disposed of, after the recovery of any useful spares. 17 June 2019.

Class 442s

## 4422411 *The Railway Children*

**THE RAILWAY CHILDREN**

2411 is seen here approaching Alexander Road footbridge, on the approach to Weymouth with a service from London Waterloo. The long sidings to the left are retained for stabling charter trains that visit the town regularly, mostly in the summer months. 13 September 2006.

It is late autumn, and the trees have just about shed their leaves for yet another year as 2411 glides past Totton with a Poole-bound semi-fast train. The freight-only line to Marchwood and Fawley diverges off the main line here behind the photographer. 29 November 2006.

Later the same day, 2411 is seen arriving at Bournemouth with a Poole to London Waterloo semi-fast service. The Bournemouth Motive Power Depot was once situated to the right of the picture and behind the train. Note one the Class 442 successors, 444017, in the siding, waiting to head to London. 29 November 2006.

Soon after leaving Poole, the unit crosses over Holes Bay, part of the large Poole Harbour with a Weymouth service. It is hard to believe now, but originally the waters of the harbour lapped right up to the line here, which was once on a raised causeway, and the bridge I am standing on did not even exist! There was also a wide-open space in the area above the train until that land was also reclaimed sometime during the 1970s. 2 January 2007.

The unit lasted until late January 2007, when it was withdrawn by South West Trains, but like many of the class, it later went on to work with Southern on Gatwick Express and Brighton services. It is seen here approaching South Croydon in the company of 2413, with just such a service. 24 July 2012.

Eventually, it was again withdrawn, this time by Southern, and yet again the unit came back to its original roots with South Western Railway. It is seen here in the company of 2414, heading south across Redbridge Causeway with a Fratton Depot to Bournemouth TRSMD crew training run. 7 February 2019.

## 4422412 *Special Olympics*

This was another unit that lasted until the final day with South West Trains – in fact, it is recorded as the very last 442 working for South West Trains, the 2105 London Waterloo to Poole on 3 February 2007. This, however, is not the first place you might have found it. Capacity for stabling units at Poole has always been quite limited, especially if one of the sidings to the west of the station was out of action, and for this reason, one of the two remaining sidings behind the station was electrified to provide extra stabling space. However, it is only occasionally used, and this was the first (and perhaps only) time a Class 442 was stabled here. 4 December 2006.

This is 2412 pausing at Basingstoke with a London Waterloo to Poole semi-fast service. 16 December 2006.

Here we see the unit approaching Worgret Junction, about a mile west of Wareham with a Weymouth-bound service. The train is approaching the point where the branch line to Swanage leaves the main line, which is just behind the photographer. The fields glimpsed through the trees to the right have an interesting history, as they were once the site of the huge World War One Wareham Camp (also known as Worgret Camp), which consisted mostly of tents and wooden huts. Training for units of the Territorial Army had been going on around the Wareham area for over 50 years, and one of the main reasons for this was because of the wide-open spaces and easy access by train. In fact, just behind the last coach of the unit, there was a short siding in use during this period, to facilitate the loading of tanks and supplies. The whole area was closed in the early 1920s and returned to agricultural use. 2 January 2007.

*Above*: The sun is now getting low as 2412 is seen on its return from Weymouth, now heading for London Waterloo once again. The train is about to cross over Rockley Viaduct, about a mile or so south of Hamworthy. 2 January 2007.

*Right*: On a damp day a week or so later, 2412 is seen again, this time at Poole station as it waits to depart with a Poole to London Waterloo semi-fast service. 13 January 2007.

I quickly turned the camera round in time to capture another of 2412's successors in the shape of 444024, which was arriving with a Weymouth service. It is interesting to note the wonky windscreen wiper on the 444, which was one of the teething problems these units encountered in the early days. 13 January 2007.

Just three years later, the unit is in service with Southern and is seen here along with 2414 approaching Clapham Junction with a London Victoria to Gatwick Airport train. This image was taken prior to the units being used on Brighton and Eastbourne services, made unmistakeable by the full 'Gatwick Express' branding. 26 February 2010.

2412 is seen coming off the Brighton line on the western approaches to Clapham Junction with a Brighton and Gatwick Airport to London Victoria service. The South Western Main Line to Southampton, Bournemouth and Weymouth is to the right of the picture. 22 October 2014.

Class 442s

# 4422413

This was one of just three units that never carried a name. Weymouth is another location that has been totally transformed since its heyday in the 1950s and 1960s. 2413 is departing for London Waterloo and passing what used to be a vast goods yard and carriage sidings to the right of picture. Behind the unit is 47488 (then being used by Fragonset), attached to four ex-Virgin Mk2 coaches, which were used as barriers when the new 'Desiro' units being commissioned at the time were being moved by diesel locos. 20 November 2003.

2413 is seen approaching Weymouth station and passing the long-disused signalbox on the right. This once large and busy box was closed back in 1987, but it was left standing for about 30 years after this due to the rear of the structure forming part of the boundary wall for the railway. It was finally demolished in the late 2010s. 2 March 2005.

We now move further north along the South Western Main Line as we see 2413 approaching Lymington Junction, just south of Brockenhurst, with a London Waterloo to Weymouth service. 2 August 2006.

A relief road built to ease the traffic in and out of Poole during the 1980s provides an excellent viewpoint to photograph trains crossing Holes Bay. 2413 approaches from Hamworthy with a London Waterloo-bound service. This view clearly shows the causeway mentioned on p 45, which was taken looking in the opposite direction from this bridge, but with the causeway infilled. 2 January 2007.

Off the Class 442s' normal route on a fine winter's day, this is 2413+2418 heading south at Chalton, just north of Rowlands Castle, with a diverted London Waterloo to Weymouth service. 9 December 2006.

2413 was withdrawn by South West Trains on the penultimate day of operation on 2 February 2007, then, after a period of storage, moved on to Southern. This is the unit passing South Croydon on a dull and dreary day, with a Brighton and Gatwick Airport to London Victoria service. 14 October 2014.

Looking very smart, and almost like brand-new units in the latest South Western Railway livery, 2413+2417 are approaching Millbrook (Hants) with a crew training run from Bournemouth TRSMD to Fratton Depot. 29 November 2019.

The pair seen in the last image are together once again as they approach the Woking stop with a Portsmouth to London Waterloo service. This was during the brief couple of months between the units entering service proper on the Portsmouth Direct line and their unexpected demise in May of this year. 7 January 2020.

## 4422414

This is another one of the units never to have carried a name. 2414 is coming around the curve on the approach to Southampton Central with a London Waterloo-bound service. 4 September 2003.

It is a great shame that this superb view of the famous Big Ben has now been ruined by the growth of the trees above the unit. Rather ironic really that, in the middle of a city, it is actually a tree that blocks a good view! 2414 has just left London Waterloo and is approaching Vauxhall station with a Weymouth service. 24 October 2003.

Another view beneath the fine roof at Bournemouth. 2414 on the left is on a London Waterloo to Weymouth service, whilst to the right is 2401 with a Wareham to Brockenhurst stopping service.
11 March 2005.

Paired up with 2409, this is 2414 leading a diverted Weymouth to London Waterloo train arriving at Havant station. The last remaining through track here has since been removed. 28 January 2006.

The unit last worked for South West Trains on 2 January 2007 and was the first unit to go to Ilford for refurbishment. However, upon its return to Bournemouth TRSMD in a plain red, white and grey livery, it was soon put into storage at Eastleigh without re-entering service with South West Trains. It is seen approaching the depot gates at Eastleigh for storage after arriving from Bournemouth together with an unidentified unit.
24 January 2007.

2414, along with 2412, was one of the first units to be refurbished at Wolverton for Southern and was hauled there on 3 December 2007. During 2008, it was eventually returned to service on the London Victoria to Gatwick and Brighton services, and it is seen here approaching South Croydon on one such service. 24 July 2012.

The unit once again found itself redundant upon withdrawal by Southern and was once again put into storage. However, it was yet again taken on by South Western Railway and was one of the first to be given the company's new livery. It is seen passing Millbrook (Hants) with a Fratton to Bournemouth TRSMD crew training run. 8 November 2018.

## Class 442s

The first refurbished 442s that South Western Railway utilised for its trains were a couple of early morning services from Poole to London Waterloo. This is 2414 (with 2406 leading) passing Freemantle footbridge and approaching Southampton Central as it heads for London Waterloo. 24 July 2019.

Back in familiar territory! This is the earlier of the two morning services from Poole, seen arriving at London Waterloo with 2414 and an unidentified unit. 13 August 2019.

## 4422415 *Mary Rose*

It is hard to believe, but many years ago there used to be quite a substantial station here at Boscombe (just north of Bournemouth), but it was closed in the late 1960s. 2415 is passing the site with a northbound test run from Bournemouth TRSMD. The unit was virtually new, having been delivered to Bournemouth just two months before. It is interesting to note that the tiny dot-matrix destination-marker blind above the corridor connection can be seen actually working; this was very rare to see, and I do not think any units actually entered service with this feature in use, as it was deemed too small for the public to see. 4 October 1988.

Not long after the opening of the Dorchester bypass, which crosses the line on the bridge seen here, this is 2415 at Syward, on the outskirts of Dorchester, with a London Waterloo to Weymouth service. This view has since become impossible owing to the inevitable tree growth in the area. 25 March 1993.

Still at Dorchester, but almost exactly eleven years later than the previous image, 2415 is getting away from South station, just out of sight behind the signalbox, with a Weymouth to London Waterloo service. 24 March 2004.

This time we are at Totton, west of Southampton, as 2415 draws to a halt with a Poole to London Waterloo semi-fast service. 3 October 2004.

The unit is captured here between Branksome and Parkstone with a Weymouth-bound service. The train is on a gradual falling gradient at this point, which steepens just after passing the bridge on which I was standing. 22 March 2005.

2415 was withdrawn by South West Trains in November 2006 and put into store before going on to work with Southern. Although a few charter trains have run over the years using 442s, they were not very common. However, this charter operated from London Waterloo using 2415+2418 for the 'Railway Children Charity Special' and was operated by the ROG, with the destination being Poole. This first image shows the train waiting to depart Waterloo, complete with a small headboard. 29 August 2016.

This second image shows the train upon arrival at Southampton Central. It was a great pity the charter could not have been extended down to Weymouth, as I am pretty sure, perhaps surprisingly, that none of these units ever made it to Weymouth again after withdrawal by South West Trains in early 2007. 29 August 2016.

Class 442s

**4422416** *Mum in a Million 1997 Doreen Scanlon*

## MUM IN A MILLION 1997 DOREEN SCANLON

This was a one-off unit, which had a different type of air conditioning system fitted and often spent periods out of service. It is seen here, in the nice evening sunlight, passing Keysworth Crossing on the straight section of line of just over two miles between Holton Heath and Wareham with a London Waterloo to Weymouth train. 30 August 2003.

Moving a little further south than the previous image, we see 2416 at Winfrith, between Moreton and Wool, with a Weymouth to London Waterloo service. Behind the photographer was the Winfrith Atomic Energy Establishment (AEE) site that was closed in 1990. All the former reactors are now shut down and are in various stages of removal. The decommissioning is ongoing, and the area is now used for other purposes. 9 July 2004.

Heading away from the camera, 2416 is on the approach to Brockenhurst with a Poole to London Waterloo service. The line in the immediate foreground is the single track to Lymington. 2 September 2004.

Under a threatening sky, 2416 is making the call at Wareham station and is about to head south with a Weymouth-bound service. 2 March 2005.

In this picture, we see the unit curving into Hamworthy to make the scheduled stop with a London Waterloo to Weymouth service. To the right is the goods-only line to Hamworthy Quay, which rarely sees any trains these days, and behind the train was where the original main line serving Wimborne and Ringwood used to join. 1 November 2006.

Heading north with an unidentified unit, 2416 is on the approach to Basingstoke, this time with a London Waterloo-bound train. The background view has recently changed quite considerably with the construction of a new block of high-rise flats. 16 December 2006.

2416 was withdrawn by South West Trains on the last day of December 2006 and was stored at Bournemouth TRSMD. The well-known location of Campbell Road, Eastleigh, is the location here as the unit is seen passing with a Bournemouth TRSMD to Eastleigh move for further storage. At the time, this was the last remaining 442 to leave Bournemouth TRSMD. 3 May 2007.

A couple of years later, 2416 was let loose from storage and in the company of 2415. The units are seen arriving back at Eastleigh after a test run up to Basingstoke. The run was handled by a Freightliner Heavy Haul driver. 2 April 2009.

Eventually taken on by Southern for Gatwick Express services, the unit is seen here approaching Battersea Park station, soon after departing London Victoria with a Gatwick Airport service. 19 May 2011.

## 4422417 *Woking Homes*

*Above*: The overall roof at a deceptively quiet looking London Waterloo is host to 2417 as it waits to depart with a Weymouth service. The platforms in the foreground are always very busy with commuter trains, but I was lucky enough to be able to grab this shot between the frequent arrivals and departures! 24 October 2003.

*Opposite above*: Much further south, at the opposite end of the South Western Main Line than the previous image, 2417 is arriving at a frosty Dorchester South station with a Weymouth to London Waterloo train. Just disappearing into the distance on the left is a rare visitor to the area in the shape of a Class 31, which was doing a route-learning run. The grassy banks seen in the top background are the Neolithic Maumbury Rings which, during the Roman occupation, more than 2,000 years after building, were adapted as an amphitheatre for the use of the citizens of Dorchester. It is now a public open space. 2 March 2005.

*Opposite below*: With the goods-only branch line to Marchwood and Fawley heading off to the left, 2417 is rounding the curve on the approach to Totton with a Weymouth to London Waterloo train. 29 November 2006.

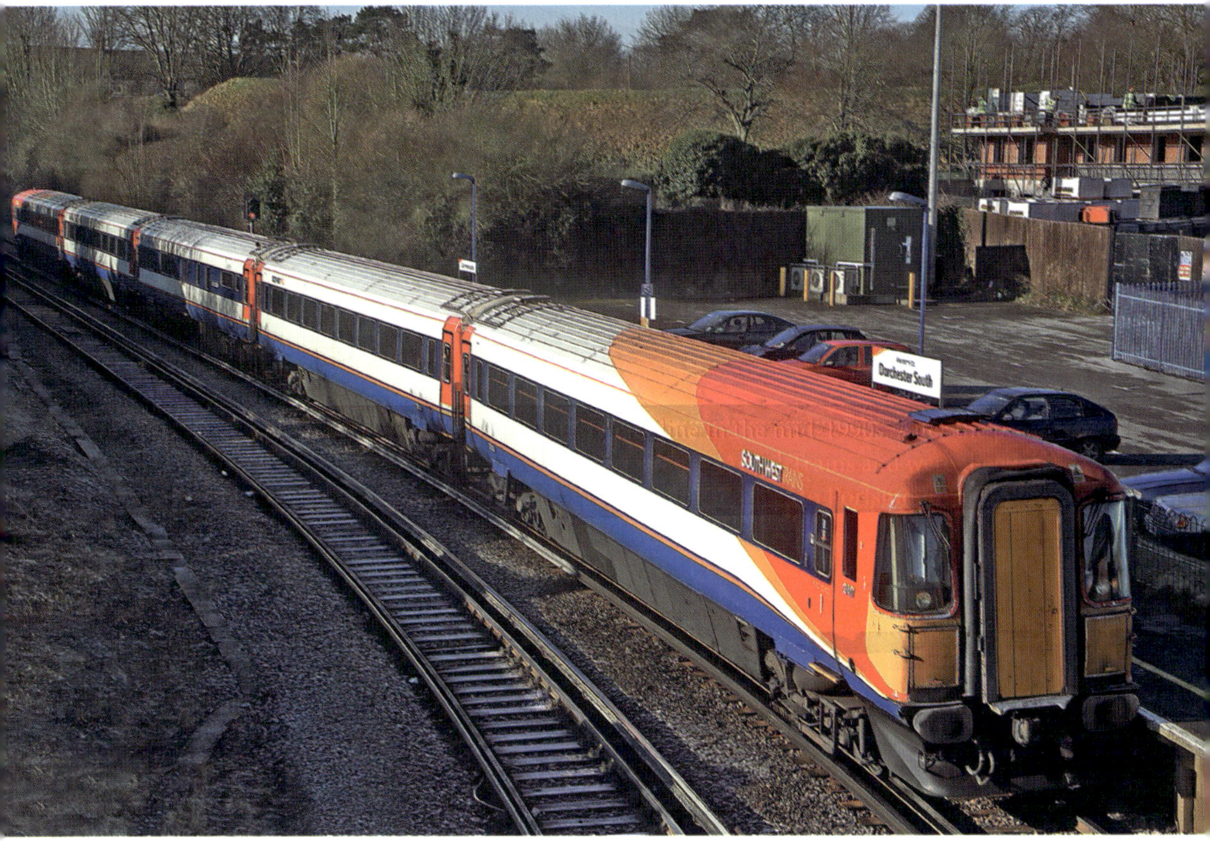
Dorchester South
SOUTHWEST TRAINS

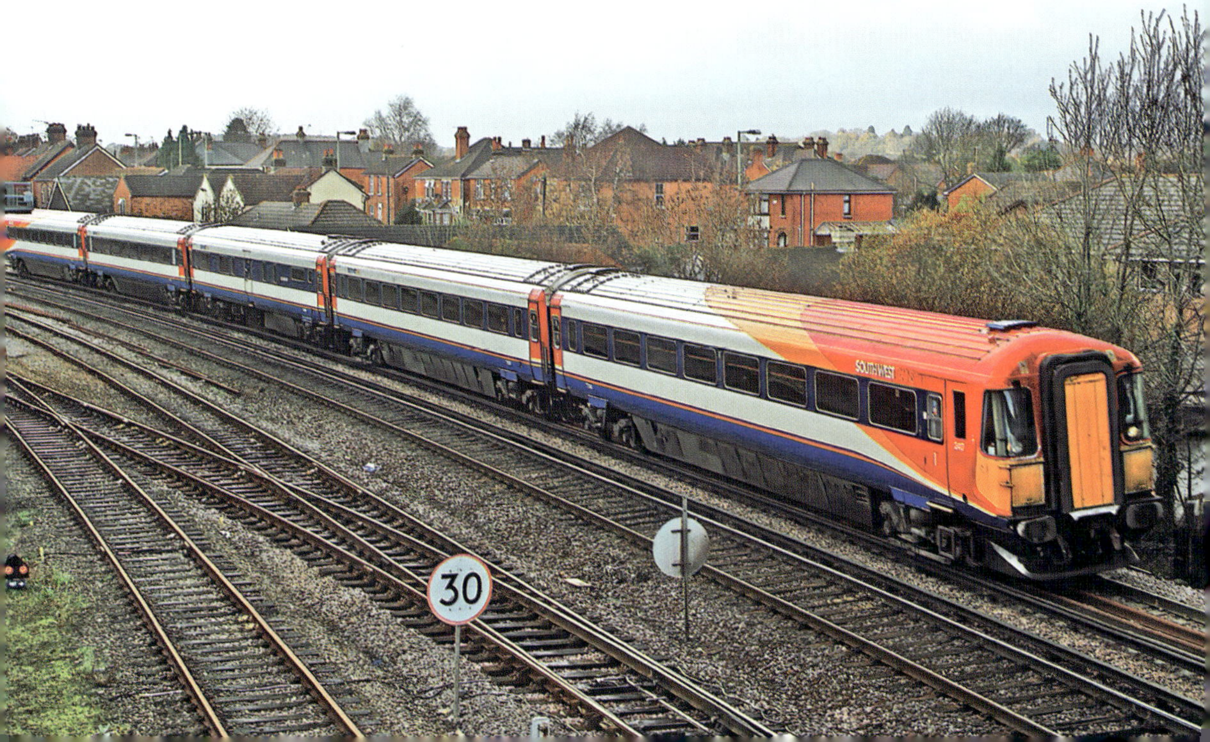
SOUTHWEST
30

## Class 442s

There is certainly no doubt about the location here as the unit awaits departure for London Waterloo. When used on Weymouth to London Waterloo services, the first-class accommodation was always intended to be leading from Weymouth, but this was not always the case. 30 November 2006.

Winding its way out of the sidings at Poole and past the signalbox, the unit is approaching the station to form a semi-fast service to London Waterloo. Since this image was taken, the signalbox here was closed and demolished when the line was re-signalled back in the mid-2010s. 4 December 2006.

Drawing to a halt at Wool, this is 2417 on a Weymouth-bound service. In the background, we can see the loading area for the sand trains that used to work from here, first operated by EWS and more recently by Freightliner. However, it has been many years now since the last one ran. 13 January 2007.

## Class 442s

Poole is another station that has undergone many changes since the late 1960s. The only thing that has not changed, however, is the very sharp curve it is situated on, seen to good effect here. 2417 is waiting to leave with a semi-fast to London Waterloo, whilst 444016, another of its successors, has just arrived in the opposite platform, working a London Waterloo to Weymouth service. 13 January 2007.

Earlier in the same day as the last image, 2417 is bringing up the rear of a terminating service from London Waterloo, arriving into Poole across the notorious level crossing. This crossing has always been a bit of a headache, even back in the days when it was a road crossing. Nowadays, it is bang in the middle of a shopping precinct and is very busy with shoppers who often try to beat the barriers when they are dropping; some even leap over them when they are down. Originally, there was another level crossing about where the fourth coach of the unit is, but this was closed back in the mid-1970s. 13 January 2007.

2417 was withdrawn by South West Trains in mid-January 2007 and put into store at Eastleigh. Another of the class that went full circle and back to where it started, 2417 is now in full South Western Railway livery passing through Millbrook (Hants), with the early winter sun starting to set, with a Bournemouth TRSMD to Fratton Depot crew training run. 29 November 2019.

During its brief time in service with South Western Railway on the Portsmouth to London Waterloo services, 2417 is seen on the rear of one of these at Woking. I could not have known it at the time, but this was the very last occasion I photographed one of these units in service. 7 January 2020.

**4422418** *Wessex Cancer Trust*

# WESSEX CANCER TRUST

This is 2418 at Baiter, just after leaving Poole with a Weymouth to London Waterloo train. In the background to the left of picture, there was once the huge Poole gasworks complex, but this was closed back in the 1970s and the whole area has since been redeveloped. 25 March 2003.

Further southwest again, and we see 2418 passing East Burton level crossing, about a mile south of Wool station, with a Weymouth to London Waterloo service. Part of the aforementioned AEE Winfrith site can be seen in the left distance. 2 December 2006.

Off route this time as 2418 is captured passing the Old Man's Beard, clinging to the bushes at Chalton, north of Rowlands Castle, with a diverted Weymouth to London Waterloo service. 9 December 2006.

Back on familiar territory at Wool, we see the unit pulling away from its station call with a Weymouth-bound service. A poor unfortunate bird appears to have got caught in the buffer beam. The level crossing here can sometimes cause very long traffic queues, especially in the summer months. 2 January 2007.

2418 was withdrawn by South West Trains at the end of January 2007 and sent for storage. This is the unit awaiting its next instalment in the works yard at Eastleigh, during the period when many units were stored here. 20 February 2009.

Its next instalment was to join Southern for the Gatwick Express services, and it is seen passing South Croydon with a London Victoria to Gatwick service. Note that, at this time, the unit carried no branding on the carriage sides. 15 October 2010.

Once Southern had dispensed with them, the unit had yet another period in store at Eastleigh, but some were occasionally given a 'warming' run in case of future use. Here we see 2418 at Basingstoke with the 5O43 10.52 Eastleigh Arlington to Basingstoke and return test run. 27 July 2016.

A month later, 2418 was out for another run from Eastleigh. It is seen here speeding through Shawford with 5O41, the 09.33 Eastleigh Arlington to Basingstoke and return, soon after departure. 22 August 2016.

A week later than the previous image and 2418+2415 are seen at London Waterloo. The pair had worked ECS from Eastleigh to London Waterloo to work 1Z42, the 09.03 ROG-operated charter to Poole. This was the first time a 442 had been seen 'on the blocks' at Waterloo for many years. 29 August 2016.

## Class 442s

### 4422419 BBC South Today

*Left*: With the driver's windscreen wiper looking to be a bit shattered, 2419 is approaching Eastleigh with a London Waterloo to Poole semi-fast service. Bizarrely, this unit was officially de-named in 2007 (although, at least one plate was missing before this date) by *South Today* presenter Sally Taylor. As far as I know, this must be the only occasion a railway vehicle has actually ever had its name officially removed. 25 March 2003.

*Below*: Gliding past a winter crop field at West Stafford, just east of Dorchester, the unit is working another Weymouth to London Waterloo train. 2 December 2006.

A rare location for any photograph is from the long-abandoned platforms of Monkton and Came Halt, between Dorchester and Upwey. This tiny halt was opened by the Great Western Railway in 1905 and closed in 1957, and up until that time existed mainly to serve a nearby golf course (which is still very much open). 2419 approaches in the low winter sunshine with a Weymouth to London Waterloo service. 2 December 2006.

2419 is waiting to depart Wareham for London Waterloo. There were originally two bay platforms at Wareham, with one ending above the cab roof and the other out of sight behind the awning on the left. Both were used by Swanage branch trains up until 1972. 4 December 2006.

The unit is seen here arriving into Southampton Central with a London Waterloo-bound service. Despite being de-named in early January, as noted earlier, the unit is still carrying a name, at least on one side. Perhaps someone simply forgot to remove it? 24 January 2007.

A few days later, 2419 is seen again at Southampton Central. This time, I took the last image from working a semi-fast service to London Waterloo. Definitely no nameplate on this side! Although this unit last worked for South West Trains at the end of January, it was not officially withdrawn until 12 February. 29 January 2007.

Another member of the fleet to find work with Southern on Gatwick Express services, it is seen here passing through East Croydon with a London Victoria to Gatwick Airport service. 12 August 2010.

### 4422420 *City of Southampton*

Pulling away from the Eastleigh call, 2420 is working a semi-fast service bound for Poole. 25 March 2003.

*Right*: With the sunshine getting low in the early winter afternoon, 2420 draws into Poole station after exiting the sidings to form a semi-fast service for London Waterloo. 29 November 2006.

*Below*: Although the unit is seen here approaching Bincombe Tunnel, between Dorchester and Weymouth, on a rising gradient, it did not really have much of an effect on these units, unlike the steeper western approach to the tunnel that did actually make them work. 2420 is heading for Weymouth with a service from London Waterloo. 2 December 2006.

*Above*: Crossing the superb Wallington Viaduct, about a mile or so west of Fareham, 2420 leads a diverted London Waterloo to Weymouth service as it heads west. I made this trip with a friend of mine, and his fine original Mini sits proudly in the foreground! 9 December 2006.

*Left*: Arriving at Platform 4 at Southampton Central, the unit is working another London Waterloo to Weymouth service. The unit was withdrawn by South West Trains just nine days later. 3 January 2007.

Twelve years later, the unit has 'come home' and, on a very wet day, the unit is seen once again at Southampton Central, now in full South Western Railway livery, working a Bournemouth TRSMD to Fratton Depot crew training run. Note that, whereas South West Trains just used the individual unit number on the cab side, Southern and later South Western Railway displayed the full class and number. 24 April 2019.

On the same day, now a little drier, 2420 is seen on the return run from Fratton Depot to Bournemouth TRSMD, this time just west of Southampton Central station. 24 April 2019.

## 4422421

*Above*: This was the third unit of the trio never to have carried a name. With some nice silver-painted buffers, the smart looking unit is seen here waiting to depart Poole with a semi-fast service to London Waterloo. Next to it is 2405, having just arrived with a Weymouth train. 9 July 2004.

*Right*: Here we see the unit approaching the stop at Totton with a London Waterloo to Poole semi-fast service. 29 November 2006.

Once the domain of Southern Railway Bulleid Pacific, King Arthur and Lord Nelson class steam locomotives, among others, 2421 is making the call at Bournemouth with the same train seen in the previous picture. This unit was one of the last to go to Ilford for an overhaul and was sent there by South West Trains just three days after this image was taken. 29 November 2006.

Upon its return from Ilford after its overhaul, this unit appeared in plain white livery, pending developments on its future. It never worked again for South West Trains and is seen stabled at Bournemouth TRSMD, where it swapped its motor bogies with another of the class. It moved away for further storage to Eastleigh at the end of this month. 7 January 2007.

Like so many, 2421 eventually found further employment with Southern on Gatwick Express services. The unit is seen here in gloomy weather, approaching South Croydon with a service for Gatwick Airport and Brighton. 14 October 2014.

Class 442s

Seen calling at Clapham Junction, during the final days the units operated with Southern, with a service from Brighton and Gatwick Airport heading for London Waterloo. 9 July 2016.

## 4422422 *Operation Overlord*

**OPERATION OVERLORD**

2422 is seen standing at Platform 3 in Southampton Central after arriving with a service from Poole. 24 October 2003.

Winding its way between the buildings, and the litter and general detritus on the northern approach to Southampton Tunnel, the unit is seen with a London Waterloo to Weymouth service. 16 March 2005.

This time we see 2422 departing Farnborough with a Poole to London Waterloo semi-fast service. Originally, there was an island platform here, but this was removed many years ago. 12 December 2006.

Withdrawn by South West Trains in early 2007, the unit went on to work for Southern. This is a chance meeting of two 442 units at South Croydon, as 2422 is working a London Victoria to Gatwick and Brighton service and passes an unidentified unit heading for London Victoria on a corresponding up service. 24 July 2012.

2422 is seen here at the well-known location of Battledown Flyover, just west of Basingstoke, with 5Z42, the 10.11 Stewart's Lane to Eastleigh Works transit move. This was crewed by ROG drivers and was one of the first workings of the class under its own power on 'home' territory for many years. 5 April 2016.

## 4422423 *County of Surrey*

With a lilac bush brightening up the overgrown cutting sides, 2423 is leading another unit past Pokesdown with a Weymouth to London Waterloo train. 26 May 2006.

Terminating trains from London Waterloo often used the up platform on arrival at Poole to save passengers having to cross the footbridge to exit the station. This is 2423 arriving with one such service. 21 October 2006.

With the moon prominent at the top of the image on a lovely winter's morning, 2423 is leading another unit across Wallington Viaduct, just west of Fareham station, with a diverted Weymouth to London Waterloo service. Although this unit ceased operations with South West Trains in early February 2007, for some reason it was not officially withdrawn by them until mid-April. 9 December 2006.

After transferring to Southern for Gatwick Express services, the unit is seen approaching South Croydon, bound for Gatwick Airport and Brighton. Note also the Class 319 passing it in the opposite direction, which was working a Thameslink service; these units are now also history and no longer work these trains. 24 July 2012.

This is 2423 on the same day as the previous image, heading back through South Croydon on its return from Brighton and Gatwick Airport to London Victoria. 24 July 2012.

## 4422424 *Gerry Newson*

The final unit of the class is seen in its early days of operation (naming was still eight years away), seen here at Winfrith with a Weymouth to London Waterloo service. 16 February 1992.

*Above*: 2424 is seen rapidly approaching Shawford station, just north of Eastleigh, with a Poole to London Waterloo semi-fast service. 19 June 2003.

*Left*: Passing the site of Lymington Junction, just south of Brockenhurst station, the unit is working another Poole to London Waterloo semi-fast service. 2 August 2006.

The transfer to Southern for Gatwick Express services took place during 2007, and the unit is seen here passing through East Croydon, heading for Gatwick Airport and Brighton. 27 August 2014.

## Class 442 'Drags'

There have been numerous times over the years when it has been necessary to haul units off the third rail, either for overhaul, diversions or storage. This section depicts a few of the many types of locomotives that have carried out these drags over the last 20 years or so.

In early 2006, the units were called in for an overhaul at Ilford. Quite appropriately, the first unit to go was 2401, and it is seen here being hauled by 47843 (then being operated by Freightliner) up the gradient on the approach to Parkstone station (Dorset). 4 January 2006.

One of the last units to go for an overhaul in late 2006 was 2421, and it is seen here about to depart Poole behind 47830, which was also working for Freightliner at that time. When it returned from overhaul at Ilford a month or so later, it never worked for South West Trains again. 4 December 2006.

The first of two images taken on the same day. This one shows 47811 (another Freightliner-operated loco at the time) hauling 2414 as 5X42, the 09.41 Bournemouth TRSMD to Ilford into Poole station, where it would continue to the sidings just west of the station to run round. 3 January 2007.

During the loco run-round in Poole sidings, I was able to hop on a train and get to Pokesdown station for this further shot of 47811+2414 powering through. Incidentally, 2414 was taken out of service the previous day and returned in a similar interim livery as 2421, never working again for South West Trains. 3 January 2007.

Another two images taken on the same day, but, unfortunately, I have no details as to where the unit was heading. This was the first occasion that one of the class had been hauled by a Freightliner Class 66/5. This is 66543 hauling 2407 at Baiter on the approach to Poole. 27 February 2005.

This is the view looking in the opposite direction from the bridge in the previous image as 66543 gets underway hauling 2407 from Poole. 27 February 2005.

The first of another two images showing 66585 *The Drax Flyer* hauling 2421+2424 away from Eastleigh, this time heading to Wolverton for refurbishment for Gatwick Express. 17 April 2008.

This is the second image I was able to obtain at Shawford of 66585 *The Drax Flyer*, powering north with 2421+2424 in tow. 17 April 2008.

On a few occasions, various members of the class were used as dead-load ECS for driver training runs behind one of the Class 73s then in use with South West Trains. The next two images show 73235 with 2423 at Millbrook (Hants) working a Bournemouth TRSMD to Southampton Central and return run. 15 April 2005.

The train is now seen heading back through Millbrook (Hants) to Bournemouth TRSMD with the Class 73 pushing. When propelling, the Class 73 can be operated from the leading cab of the unit. 15 April 2005.

A few days prior to the working seen on p 89, this is 73109 *Battle of Britain 50th Anniversary* departing Poole in a snow flurry, hauling another unidentified working with an unknown Class 442 northbound. 22 February 2005.

73109 *Battle of Britain 50th Anniversary* is seen again, this time at Brockenhurst on a damp and miserable day, hauling 2422. On this day, a Class 73 was required in the London area, and to avoid one being sent light engine, it grabbed a ride from Bournemouth on the front of the 12.00 Weymouth to London Waterloo service, thus providing the rare sight of a loco-hauled train to Waterloo. The footbridge seen in this view has since been replaced by a modern structure. 22 November 2006.

This is a crew training run operated by Southern, prior to the class entering service on the Gatwick Airport runs. 2424 is propelled westbound towards South Croydon by 73204. 13 September 2007.

During 2010, Southern took on a couple of extra units, as their existing fleet was being used on various other services to Eastbourne etc. Although repainted, including the undercarriage and bogies, 2422 has a few of its windows boarded over. It is sandwiched between yellow-liveried 73212 and another unidentified Class 73 as they approach Tonbridge after a test run to Ashford and back. 2 March 2010.

*Above*: A couple of weeks after completing the test runs seen in the previous image, this is silver-liveried DB-operated 67029 hauling 2422 across Battersea Bridge, soon after departure from Stewart's Lane to Wolverton for overhaul and for further use with Southern. 18 March 2010

*Right*: During their time in storage at Eastleigh, various units were moved between there and Bournemouth TRSMD. This is ROG 37608+2410 at Mount Pleasant crossing, near St Denys, with 5Y85, the 12.20 Eastleigh Arlington to Bournemouth TRSMD. 1 February 2018.

This is ROG 37884+2403 approaching Farnborough (Main) with 5O37, the 09.02 Stewart's Lane TRSMD to Eastleigh Arlington. This unit had just been withdrawn by Southern and is heading for further storage, prior to being brought back into service for South Western Railway a few years later. 12 August 2016.

During the mid-2010s, units seemed to come and go from all over the place, seemingly just for the sake of it! This is, again, ROG-operated 37884+2411 approaching Basingstoke with 5O86, the 12.49 Three Bridges to Eastleigh Arlington. Similar to 2403, this unit was later taken on by South Western Railway. 9 November 2016.

It was unusual for any of these storage runs to Ely to be top and tailed by two locomotives, but this is 47812, 2423 and 37884 approaching Hook with 5E42, the 12.49 Eastleigh Arlington to Ely Papworth Sidings. Use of a long zoom lens meant I was just about able to get a shot before that Class 159 got in the way! I think the only reason for the top and tailing on this occasion was to get an extra loco over to Ely. 17 August 2016.

A couple of weeks later and there was no such problem with the Class 159, with just one loco hauling. This is 47815+2420, again approaching Hook with 5E42, the 12.45 Eastleigh Arlington to Ely Papworth Sidings. At the time, ROG had just taken on a few former Virgin Class 47/8s to help out with these drags. 31 August 2016.

Another two weeks later, and the same locomotive was used again for a similar move. This is 47815+2415+2416 passing Basingstoke with 5E42, the 12.51 Eastleigh Arlington to Ely Papworth Sidings. This image was taken from a convenient multistorey car park. 14 September 2016.

Another Class 47, operated at the time by ROG, this is 47813 *Jack Frost* (with 47815, now named *Lost Boys 68-88*, on the rear) approaching Eastleigh hauling 2422 as 5Q87, the 09.40 Bournemouth TRSMD to Eastleigh Works. I am not really sure why the unit could not have worked under its own power, but I was not complaining! 17 September 2019.

Of all the loco types that have hauled any of the units, this is the only time, as far as I am aware, that a Class 31 was used. Hired in by ROG at the time, DCR-liveried 31452 is hauling 2405 past Basingstoke with 5L46, the 12.33 Eastleigh Arlington to Ely Papworth Sidings, where the unit would be stored. Upon their withdrawal by Southern, many units were first moved to Eastleigh and then to Ely Papworth Sidings for further storage, as room at Eastleigh was becoming an issue. 21 October 2016.

*Above*: This is 73141+2409 approaching Paddock Wood with a return test run to Chart Leacon Depot that had run to Ashford. 2409 was originally intended to be a source of spare parts for other units, but it was decided to get it back in service during 2010 for Southern, as 442s were now also being used on Eastbourne services. 23 July 2010.

*Left*: With the Class 73 having run round, the same test train is seen again, this time arriving at Tonbridge, with the Class 73 now propelling. The ensemble made a few trips during the day to check all was well with the unit. 23 July 2010.

**Further reading from**

As Europe's leading transport publisher, we produce a wide range of railway magazines and bookazines.

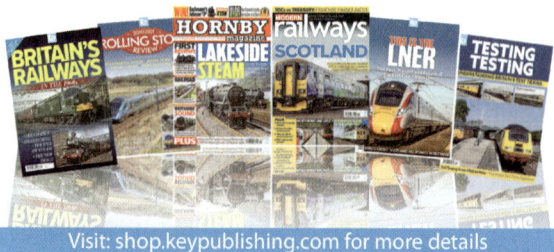

Visit: shop.keypublishing.com for more details